21st Century Junior Library

UNDERSTAN EXTREME TEMPERATURES

Responding to Natural Disasters

Olivia Williams

Published in the United States of America by:

Cherry Lake Press
2395 South Huron Parkway, Suite 200, Ann Arbor, Michigan 48104
www.cherrylakepress.com

Reading Adviser: Beth Walker Gambro, MS, Ed., Reading Consultant, Yorkville, IL
Photo Credits: © John McCormick/Shutterstock.com, cover, 1; © johnnychaos/Shutterstock.com, 5; © Pat Tr/Shutterstock.com, 6; © Katiekk/Shutterstock.com, 7; © vitrolphoto/Shutterstock.com, 9; © Mr.Somchai Sukkasem/Shutterstock.com, 10, 11; © Tanja Tatic/Shutterstock.com, 12 [left]; © J.J. Gouin/Shutterstock.com, 12 [right]; © Chayathorn Lertpanyaroj/Shutterstock.com, 13 [top]; © pondpony/Shutterstock.com, 13 [bottom, left]; © Ezzolo/Shutterstock.com, 13 [bottom, right]; © outdoorsman/Shutterstock.com, 15 [top]; © Gilles Brignardello/Shutterstock.com, 15 [bottom]; © TR STOK/Shutterstock.com, 16; © Sean Leonard/Shutterstock.com, 19; © Proxima13/Shutterstock.com, 20

Cherry Lake Press is an imprint of Cherry Lake Publishing Group.

Library of Congress Cataloging-in-Publication Data
Names: Williams, Olivia (Writer of children's books), author.
Title: Understanding extreme temperatures / Olivia Williams.
Description: Ann Arbor, MI : Cherry Lake Publishing, [2022] | Series: Responding to natural disasters | Audience: Grades 2-3
Identifiers: LCCN 2021043048 | ISBN 9781534199200 (hardcover) | ISBN 9781668900345 (paperback) | ISBN 9781668901786 (pdf) | ISBN 9781668906101 (ebook)
Subjects: LCSH: Temperature—Juvenile literature.
Classification: LCC QC271.4 .W55 2022 | DDC 536/.5—dc23/eng/20211005
LC record available at https://lccn.loc.gov/2021043048

Cherry Lake Press would like to acknowledge the work of the Partnership for 21st Century Learning, a Network of Battelle for Kids. Please visit http://www.battelleforkids.org/networks/p21 for more information.

Printed in the United States of America
Corporate Graphics

ABOUT THE AUTHOR

Dr. Olivia Williams is from mid-Wales in the United Kingdom. She has traveled to many countries to help people after natural disasters. She has felt earthquakes and aftershocks, and seen damage from floods, typhoons, and wildfires. The natural environment can be harsh but also beautiful. Olivia loves to hike and run in the mountains and photograph the incredible things she sees.

Cherry Lake Press

CONTENTS

WHAT ARE EXTREME TEMPERATURES?

Temperatures are different all around the world. Some places are cold. Others are hot. Some places are somewhere in the middle. Scientists record average temperatures everywhere. They use this information to determine whether regions are getting colder or hotter.

Extreme temperatures occur when there is a **significant** change in temperature. Scientists sometimes call these changes a "heat wave"

Countries by the equator experience the least variation in temperature and seasons.

or "cold wave." When temperatures rise by 10 degrees Fahrenheit (–12 degrees Celsius) for a few days, it is called a heat wave. When temperatures fall rapidly, it is called a cold wave. England was the first country to keep a running

Death Valley, California, is said to be one of the hottest places on Earth.

The coldest place on Earth is located in Antarctica.

record of temperature. It started its record keeping in 1659.

In the late 1700s to mid-1800s, transporting goods and traveling became easier. Steamboats and trains made travel faster. This period of time is called the Industrial Revolution. Before the Industrial Revolution, people relied heavily on animals and boats. During the Industrial Revolution, people started using wood to power machines. They also used **fossil fuels** like coal, oil, and natural gas.

Fossil fuels still help power many of our machines today. Burning fossil fuels for energy releases gases into the air. The gases form a thick blanket around Earth. This gas layer makes it difficult for heat to escape Earth's surface. The trapped heat makes Earth hotter. Scientists call this the **greenhouse effect**. The gases are called **greenhouse gases**.

People have used fossil fuels for hundreds of years. We use them for heating and cooking. We use them to power our cars. We also use them to make personal care products, plastic, and household cleaners.

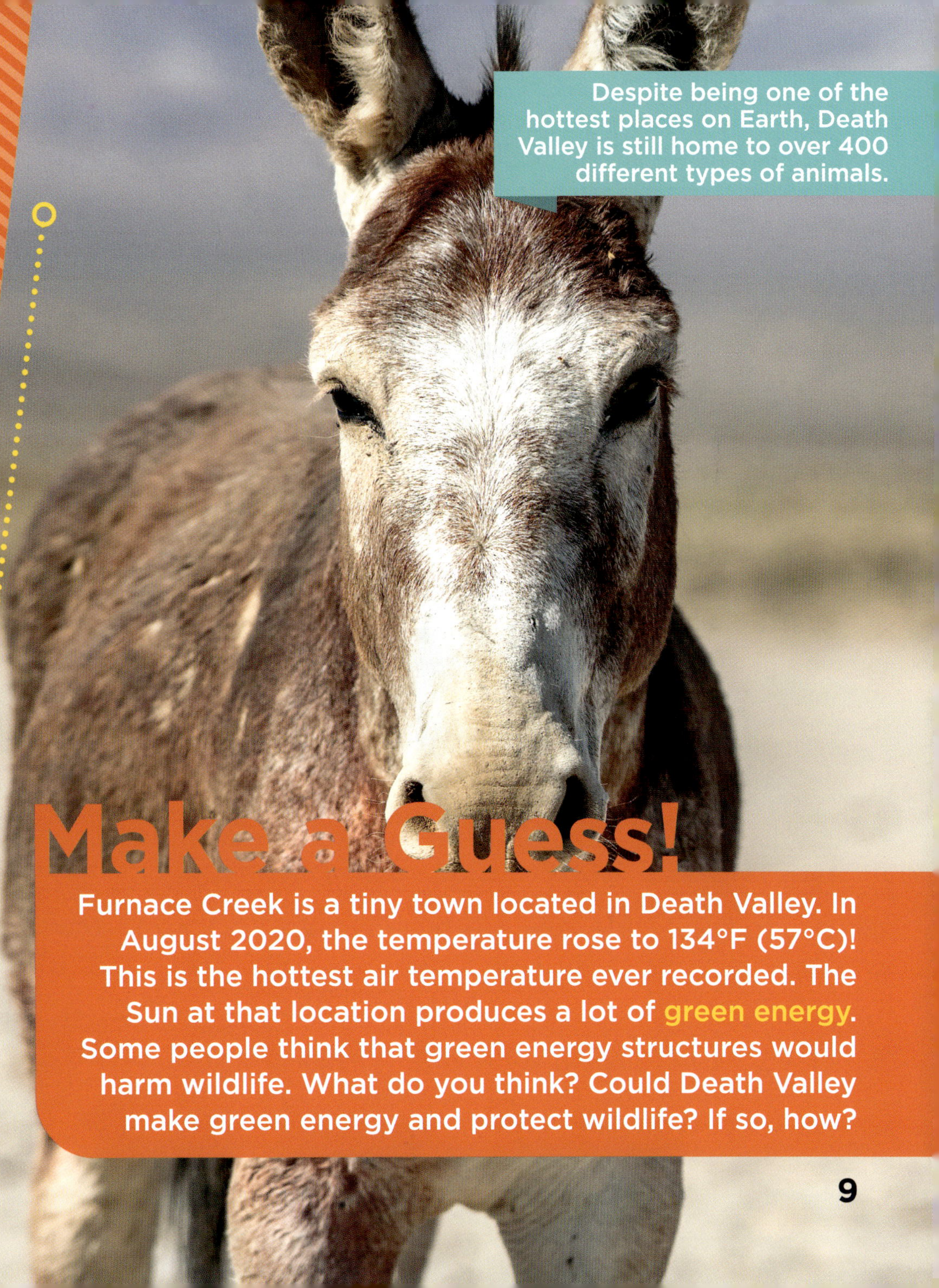

Despite being one of the hottest places on Earth, Death Valley is still home to over 400 different types of animals.

Make a Guess!

Furnace Creek is a tiny town located in Death Valley. In August 2020, the temperature rose to 134°F (57°C)! This is the hottest air temperature ever recorded. The Sun at that location produces a lot of **green energy**. Some people think that green energy structures would harm wildlife. What do you think? Could Death Valley make green energy and protect wildlife? If so, how?

HOW EXTREME TEMPERATURES AFFECT THE PLANET AND PEOPLE

Earth has regions of extreme cold. But it is extreme heat that causes problems. Extreme heat worsens **natural disasters**. It increases their frequency and intensity. It affects typhoons, tornadoes, wildfires, and flash floods. It also causes rain patterns to change. This can create natural hazards like **droughts**. Droughts can last so long that they make it impossible

for people to live and work. Without water, people struggle to care for livestock and grow food. These events can lead to human **migration**. When people migrate, conflict can occur. Conflict creates tension, stress, and sometimes even wars.

Extreme heat also causes diseases to spread more easily. Deadly diseases like malaria and Ebola thrive in high temperatures. Extreme heat also has a bad effect

Stay hydrated and keep cool! Extreme heat can lead to deadly heat strokes. Warning signs of a heat stroke include rapid heartbeat and confusion.

on people's physical and mental health. Extreme heat puts stress on the things that people rely upon. Power grids stop working and cause electricity blackouts. Airplanes may be unable to fly. Roads melt and railroads buckle, making travel dangerous and sometimes impossible.

Extreme heat also threatens some of Earth's most beautiful and mysterious environments. **Glaciers** and sea ice melt as greenhouse gases warm seas and oceans.

Think!

Ethiopia is a country in eastern Africa. This country is very badly affected by droughts. Droughts are caused by climate change. We can combat this by acting responsibly and using fewer fossil fuels. What do you think is the best way to encourage people to change their habits?

Polar bear, seal, and penguin habitats disappear. Animals find it hard to find food. The water also becomes more acidic as seawater temperatures rise. Acidic water harms smaller sea creatures and destroys their shells. It also destroys coral reefs and endangers the **biodiversity** of the oceans.

Earth naturally goes through changes in climate over long periods of time. In fact, it has experienced several ice ages and warming periods. But human activity is speeding up this naturally occurring change.

PROTECTING AGAINST AND RECOVERING FROM EXTREME TEMPERATURES

The rising temperature of the planet is called climate change. Climate change is an emergency. But it can be managed by people changing their habits. The more we use sustainable energies, the healthier the planet will be. Sustainable energies use naturally occurring **phenomena** without harming them. Sustainable energy can come from sources such as the Sun, wind, water, and tides.

Replanting forests that were destroyed by **deforestation** also helps fight climate change. Some forests are being left to regenerate naturally. This helps bring wild animals and vegetation back to these landscapes.

Educating people also plays an important role in reversing climate change. We can learn about daily changes we can make to slow the increase of temperatures. We can reuse shopping bags and recycle plastic, paper, and cans. We can also buy

Create!

Everyone has a routine. You have a routine when you get ready for school. You have a routine when you get ready for bed. Think of ways you can incorporate "greener" habits into your routine. Maybe use the Sun's natural light instead of turning on a lamp. Carpool with a friend. Create a list of things you can do. Brainstorm a few ideas with your family or friends!

Despite the quickly increasing temperature of Earth and shorter winters, scientists believe people will experience more extreme blizzards.

Many cities and people use special salt to clear ice from sidewalks and roads. But this is sometimes bad for the environment. Instead, environmentalists suggest using pickle or beet juice!

locally grown food or grow our own! Remember to turn off lights when you leave a room, and eat less meat if you can. Email politicians and companies and ask them to do more for the planet.

Ask Questions!

There are groups all over the world doing great things for the climate. Go online and see what's being done in your state. In your opinion, is your state doing enough? What else would you like to see your state doing to protect Earth?

ACTIVITY

Save the Environment

Did you know you can help fight climate change? One way you can help is by saving water. Below is a list of a few things you can do to help the environment. Type out and print this list or write it on a piece of paper. Tape it in a few places around your home to act as a reminder for you and your family!

- Use a reusable bottle instead of buying bottled water.
- Turn off faucets when not in use.
- Take shorter and colder showers.
- Reuse water.
- Run the dishwasher and washing machine only when full.
- Sweep the driveway instead of using a hose.

GLOSSARY

biodiversity (bye-oh-duh-VUHR-suh-tee) the variety of life and living creatures on Earth, in all its forms and all its interactions

deforestation (dee-for-uh-STEY-shuhn) the action or process of clearing forests

droughts (DROWTZ) long periods of little or no rain

fossil fuels (FAH-suhl FYOOLS) fuels made from very old decomposing plants and animals; these fuels include coal, oil, and natural gas

glaciers (GLAY-shuhrz) slowly moving masses of ice formed by compacted snow

green energy (GREEN EH-nuhr-jee) energy that is generated from natural resources, such as sunlight, wind, or water

greenhouse effect (GREEN-hows ih-FEKT) process that occurs when heat from the sun is trapped under a blanket of gases in Earth's atmosphere

greenhouse gases (GREEN-howss GAH-suhz) gases that absorb heat, including carbon dioxide, methane, ozone, nitrous oxide, and fluorinated gases

migration (mye-GRAY-shuhn) movement of people from one place to another

natural disasters (NAH-chuh-ruhl dih-ZAH-stuhrz) natural events such as floods, earthquakes, or hurricanes that cause great damage

phenomena (fi-NOM-uh-nuh) observable facts or events

significant (sig-NIH-fih-kuhnt) an important or large change that could change our experience or understanding of something

FIND OUT MORE

Books

Drimmer, Stephanie Warren. *Ultimate Weatherpedia: The Most Complete Weather Reference Ever.* Washington, D.C.: National Geographic, 2019.

Swift, Keilly. *How to Make a Better World: For Every Kid Who Wants to Make a Difference.* New York, NY: Dorling Kindersley, 2020.

Website

Going, Going ... Gone?

https://www.amnh.org/explore/ology/biodiversity/going-going-gone
Extreme temperatures also affect wild animals. Read more about endangered and extinct animals.

INDEX